All information is from reliable sources and studies by credible scientists

All rights reserved ©

introduction

Welcome to "Interesting Facts for Curious Minds", a book designed to spark your imagination and feed your curiosity! Whether you're a child, teenager, or adult, there's always something new and fascinating to learn about the world around us.
Have you ever wondered how far away the stars are, how your body heals itself, or how ancient civilizations shaped the world we live in today? This book takes you onan exciting journey through a varietyof topics, from the wonders of space and nature to the mysteries of history and technology.

In each section, you'll find amazing facts that are not only fun to know but also help you understand the extraordinary systems and events that make our world so remarkable. Whether you're looking to impress your friends with unusual facts or simply explore something new, this book has something for everyone.

So, grab a seat, open your mind, and get ready to dive into the incredible, surprising, and sometimes downright strange world of knowledge. The more you learn, the more curious you become!

Let's get started on this adventure of discovery!

Astronomy and Space Science

Cosmic cemeteries

tThe universe is a living organism with life and death and graveyards where stars are buried. Let's consider the new scientific discovery of the collapse of stars inside supermassive holes Scientists have discovered evidence of a star or planet falling into a black hole in the interior of our galaxy. Scientists say that this hole near the center of our galaxy strongly attracts the star's material and swallows it, sucking it up like a vacuum cleaner sucks up dust. Scientists observed a sudden flow of a stream of gas from the star towards an unknown direction and forming a vortex around the black hole, knowing that the black hole cannot be seen, but the scientists saw the vortex formed around it as it swallows the star.

At the heart of this galaxy of more than 100 billion stars lies a supermassive black hole, swallowing everything in its path as it moves through space at great speeds. It's not the only one. There are millions, perhaps billions, of them scattered throughout the vast universe.

The echoing light from the swallowing process is being detected by NASA scientists, who confirm that the black hole exists and is consuming a lot of planets, stars, dust... and sometimes even some stones. They are all "food" for this hole... Hallelujah!

This is a drawing of the process of a star being swallowed by a black hole, the black hole in the center and around it is a whirlwind of gas where the star does not directly reach the inside of the hole, but the star's matter turns into gas first due to its enormous gravity, and then it is swallowed by the black hole.

A black hole weighs a billion times the weight of the Sun. The material of the swallowed star is the fuel for this hole that cleans up the universe. Therefore, it is considered a giant cosmic graveyard!
A star at the end of its life 4200 light years away, this is the end of stars, appearing in bright colors. with a surface temperature of 50,000 degrees Celsius. These colors are caused by the gases flowing out as the star collapses. The wavelengths of the light from the explosion vary depending on the nature and composition of the gases.

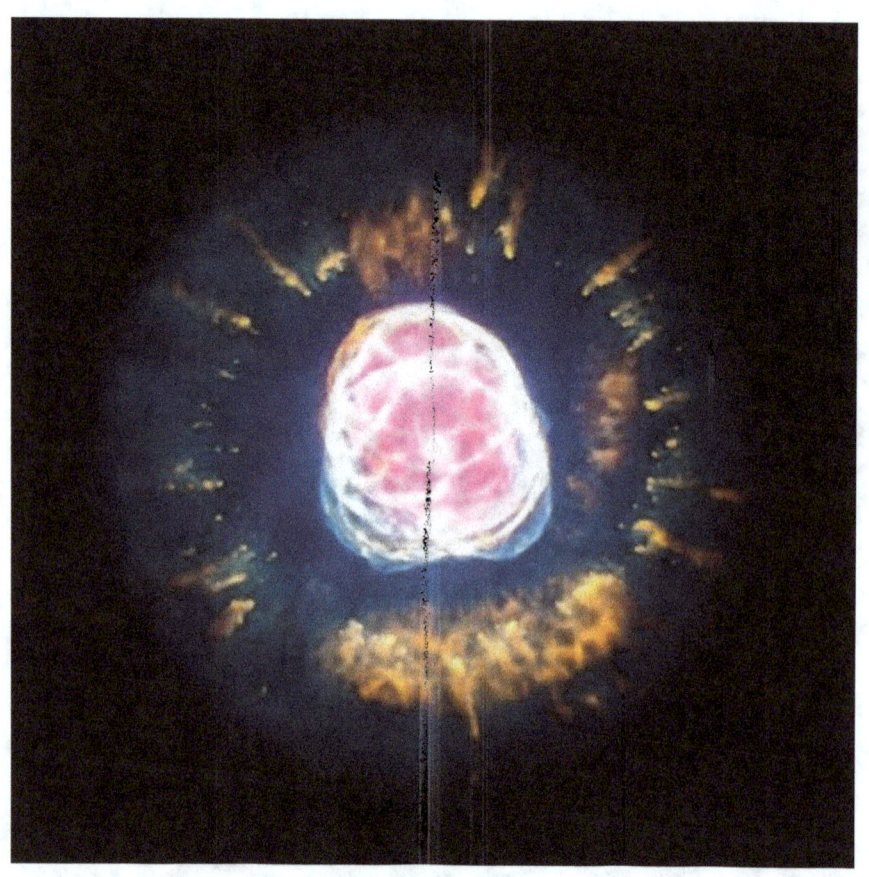

References

UPI.Science_News/ 2013/10/24/Echoes- of-light-said- evidence-of-black- hole-consuming- star-or-planet

cbc.ca/news/ technology/if-you- re-a-hungry-black- hole-try-snacking- on-a-star

Water is everywhere in the universe

Until now, there have been multiple theories about the origin of the universe and the origin of water. But recent research confirms that water is uniformly distributed throughout the universe, which baffles scientists. Let's consider Scientists have found water molecules in meteorites coming to Earth, and have come to the conclusion that water exists in space and was formed billions of years ago, and that water is not specific to the planet Earth but is widespread in the universe.
God(ALLAH) indicated the existence of water after the creation of the universe in QURAN.

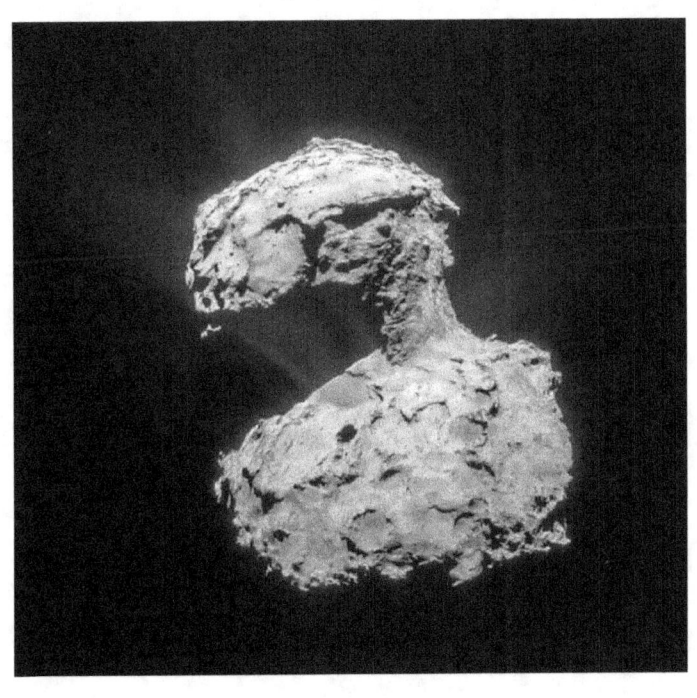

Studies of meteorites and comets conducted by NASA show that there are billions of these comets and meteorites full of water swimming in the depths of the universe, and scientists are puzzled to find an explanationfor this. When and where did these huge quantities of water form? The answer may be that water was formed soon after the universe was created and spread everywhere with the expansion of the universe

Therefore, the presence of water everywhere in the universe means that water was not formed in a specific place but everywhere according to a strict order. Therefore, the Qur'an is in complete agreement with scientific observations, as the Qur'an links the creation of the universe to the creation of water. And science confirms that water is everywhere, which indicates that it was formed early and billions of years ago.

References

nasa.gov/feature/rosettas-miro-instrument-maps-comet-water

Planet Hell

There are many blessings around us that we don't feel, but when we look and meditate on these heavens, we realize the value of these blessings.
One of God's mercies to us is that he made us live on an earth that is 150 million kilometers away from the sun, this sun is verynecessary for our life and without the sun,life on earth would not continue. But whatdoes that mean? What if our earth is close to the sun?

If the earth is close to the sun, it will overheat and end life on earth. Therefore, God Almighty has set examples in this universe for us to realize the Creator's blessing on us. One such example is a planet whose temperature rises to more than 2,400 degrees Celsius due to its proximity to the star it orbits.

An imaginary image of the planet Cancri 55E, where the surface temperature at night is more than 1,100 degrees Celsius. and by day it's 2,400 degrees Celsius. It's like a metal melting furnace!

This planet orbits its star or sun at a rate of one revolution every 18 hours. So a year for this fiery planet is only 18 hours... and it travels a distance of more than two million kilometers.

References
slate.com/blogs/bad_astronomy/2016/03/31

Cosmic factories

Every day there's new research confirming the existence of life in the universe, even if it's regularly distributed. The University of Kent is one of them. Let's contemplate
 In research from Imperial College London and the University of Kent, researchers have found that the universe has factories for the building blocks of life, called amino acids. They found that when an ice meteorite hits a planet, amino acids spontaneously form on the surface of the meteorite.

Researcher Dr. Zita Martins says:
"Our work shows that the basic building blocks of life can be assembled anywhere in the Solar System and perhaps beyond ".
 The research, published in the journal Nature Geoscience, shows that there is a regular distribution of the molecules of life (organic molecules are the basis of living matter), which are spread across the universe in an astonishing . way.
The new research proves that there are cosmic factories for the production of life scattered everywhere, both in the solar system and beyond.

The Qur'an refers to the existence of diffuse life in the heavens and the earth

References

Scientists discover cosmic factory for making building blocks of life

Discovering the Earth's diminishing limbs

The Earth is continually shrinking at its edges. This is new evidence from the American Geological Society that a continent has disappeared near Australia, meaning that the land area has decreased by the amount of water that has flooded

After long studies of the Earth from satellites, it turns out that the Earth's crust is made up of a set of plates. These plates are in constant motion. The edges of these plates are constantly being eroded by climate and sea currents.

A tiny fraction of the atmosphere is constantly escaping out of the Earth, but we hardly feel it, but accurate measurements show that it is decreasing... It's also decreasing from the edges of the continents by flooding them with water. This was revealed by scientists some time ago.

One of the recent discoveries is the discovery of a new continent three-quarters the size of Australia, and they called it Zeelandia. But it was submerged in the ocean millions of years ago and only two islands remain, including New Zealand. This continent was formed tens of millions of years ago and has an area of 4.5 million kilometers. 94% of it was submerged, leaving only two prominent islands, New Zealand and New Caledonia.

References

nbcnews.com/news/world/scientists-say-they-ve-discovered-hidden-continent-under-new-zealand-n722796

The earth is breathing

Scientists have studied the Earth's global system of carbon change and found that the Earth is like a living organism with lungs to breathe with

According to National Geographic, our Earth contains at least 3.1 trillion trees, and each tree has tens of thousands of leaves. For example, a large oak tree has up to half a million leaves. The strange thing is that each leaf is equipped with hundreds of thousands of respiratory openings called stomata. Each opening acts like a lung, taking in carbon and releasing oxygen.

So if we take into account grasses, weeds, and various types of plants, we are facing a huge number of more than a trillion trillion trillion openings (living lungs) working together on this planet in a coordinated system. They absorb enormous amounts of CO_2 and expel enormous amounts of O_2, so we have a fully breathing planet with a very precise system.

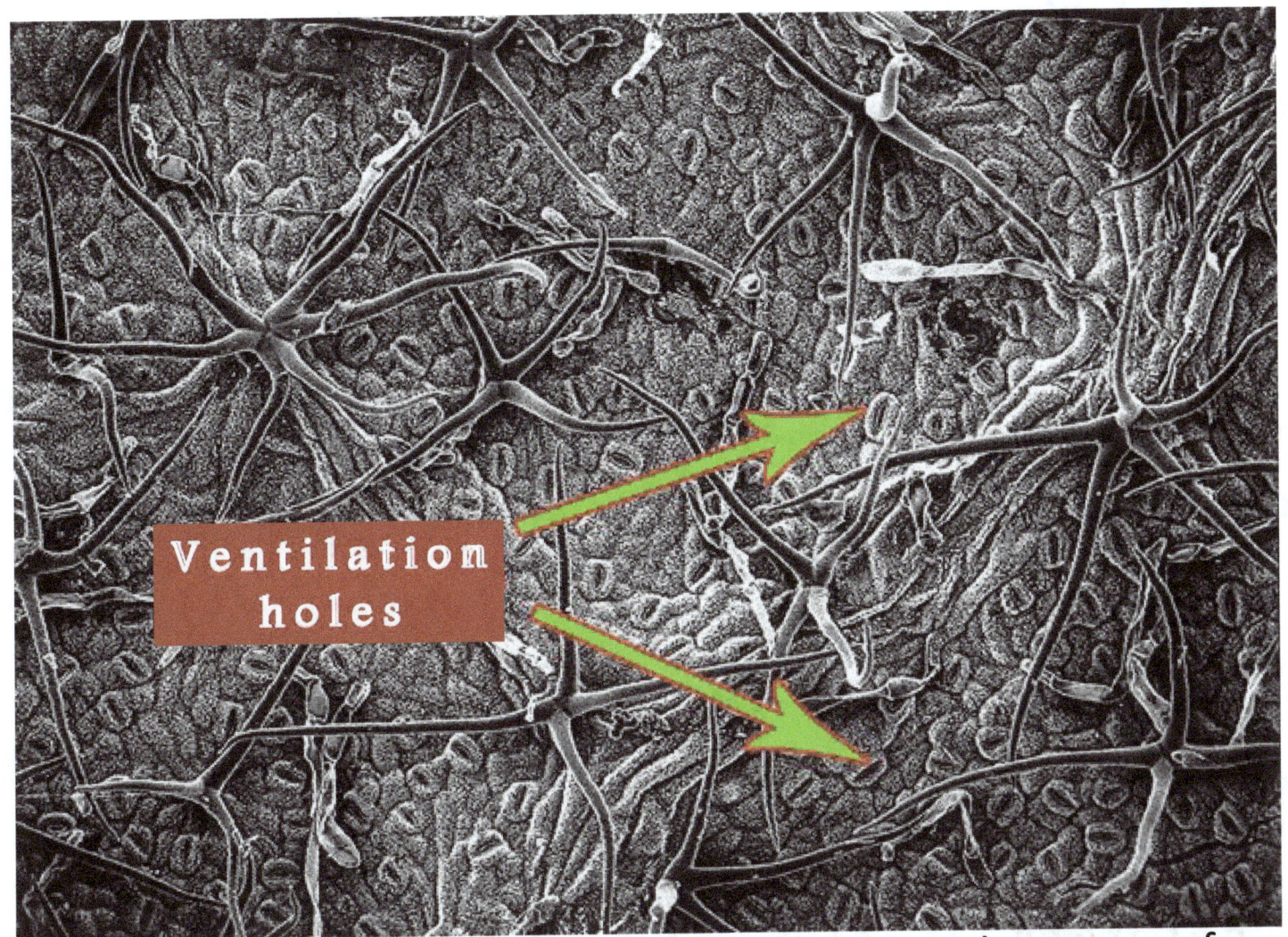

A scanning electron microscope image of a plant leaf showing hundreds of respiratory openings, where each opening acts like a real lung, taking in carbon gas and releasing oxygen gas and contributing to the ecological balance.

In the warm months, the leaves grow at about the same time and begin to absorb carbon and release oxygen. Then, in the colder months, the proportion of carbon gas increases

In other words, trillions of trees operate according to a well-organized annual cycle, which means that trillions of trees work according to a fixed global system... Without this system, there would be no life on Earth.

Without the presence of seas on Earth in exactly this proportion, the Earth would not be what it is today. Seas are essential for the emergence and continuation of life, and seas contribute to purifying the atmosphere and maintaining the ecological balance through the carbon cycle. Seas are important for rainfall for plants. Seas are important for maintaining the atmosphere. They are important for living organisms. Therefore, it is impossible to believe that these seas arose as a result of random evolutionary processes of the Earth's crust.

Scientists have also found that this global system of Earth's respiration involves the seas and other living organisms. They found that the rain cycle, or the so-called water cycle, contributes significantly to maintaining the Earth's atmosphere and maintaining the ecological balance

This water cycle works with plants, oceans and other living things in a harmonious and integrated system... There must be someone who oversees this system. Things can't just happen randomly.

References
The Earth Has Lungs. Watch Them Breathe

Discovery of an electronic roof that preserves the Earth

It's a very strong wall of electrons and protons that envelops and protects the Earth and acts like an insulating roof.

The Earth is fraught with danger from every direction. Scientists have found that massive amounts of deadly electrons traveling at close to the speed of light surrounding the Earth at a distance of more than 10,000 kilometers. These electrons, if they reached the Earth, would cause many problems for living beings.

If the process was done according to coincidences, randomness and the laws of evolution, life on Earth would have collapsed because of these electrons. But what surprised scientists is the existence of a very strong and invisible roof or shield that surrounds the earth and protects it from the evil of these destructive electrons.

Scientists have discovered an invisible shield about 11,584 kilometers away from the planet Earth, which protects the planet from dangerous super-fast electrons, which travel in space at a speed close to the speed of light.

The inner belt closest to the Earth is located between 1000 to 7000 kilometers away and consists mostly of protons with an average energy of 30 million electron volts, while the outer belt is located between 10,000 to 40,000 kilometers away and contains mostly electrons with an average energy of 1 million electron volts.

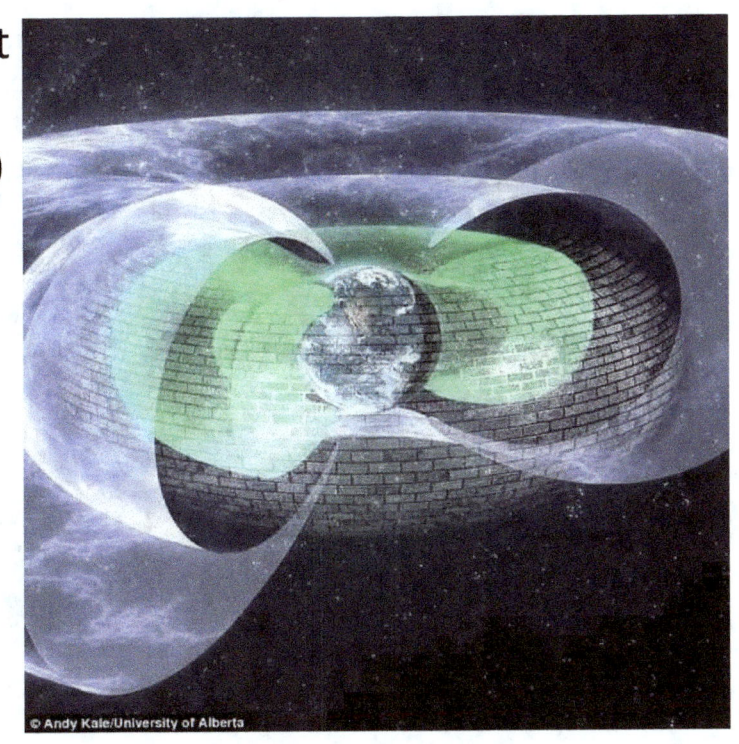

During the research, scientists discovered an "incredible" layer on the inner edge of the outer belt, about 11,584 kilometers from Earth, that looks like a glass wall or a protective shield that prevents ultrafast electrons from moving towards the Earth's atmosphere.

Professor Daniel Baker of the University of Colorado Boulder emphasizes that this shield is like a glass barrier that protects the Earth from deadly electrons, which, if they reached the Earth, would change the climate, disrupt power plants and increase cancer rates.

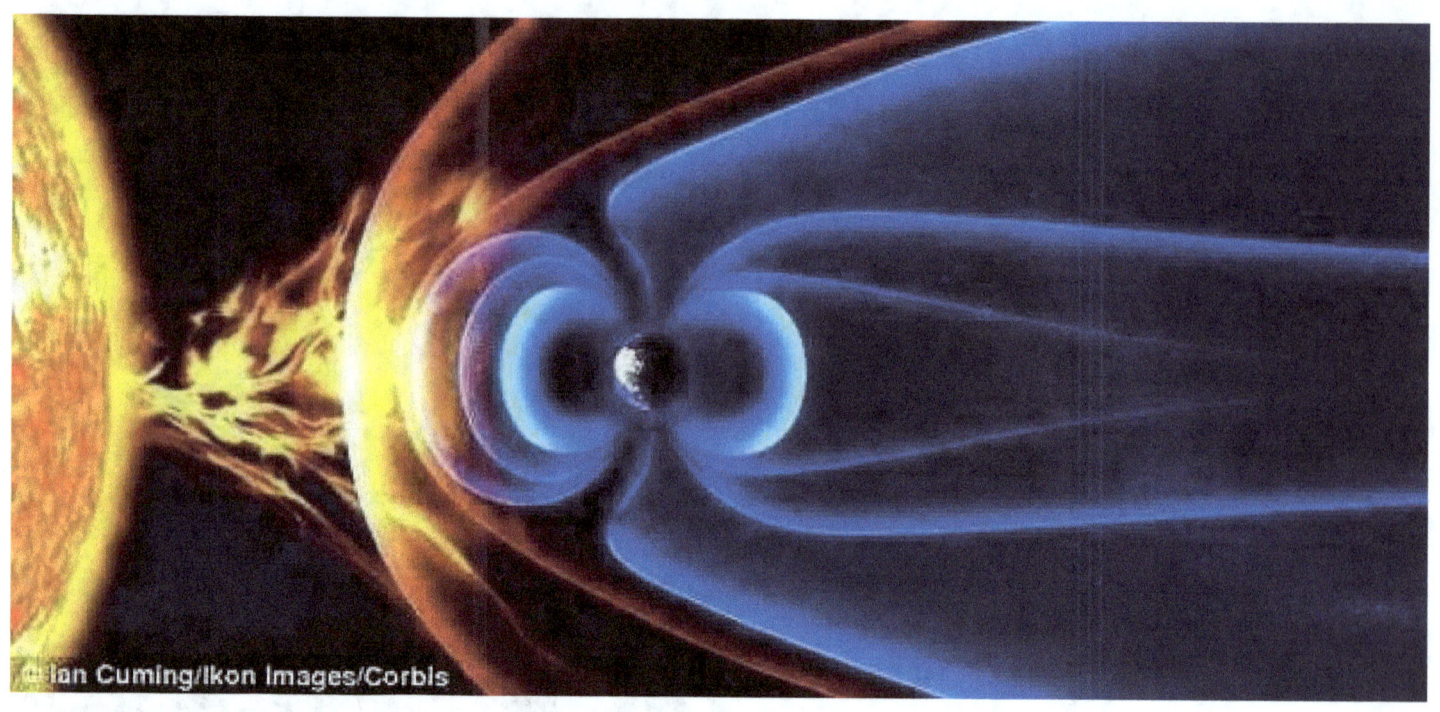

This drawing represents the Earth in the sun's orbit and we observe the Earth's strong magnetic field repelling the attack of cosmic rays and destructive solar winds
  This shield creates electromagnetic waves that scatter deadly electrons away from the Earth, and protects our planet from various cosmic particles. Although we don't feel any danger to us, dangers are surrounding the Earth from all sides without us even realizing it!
  The sun is a major source of electrons and killer rays (solar wind), and if it weren't for the atmosphere, strong electromagnetic belts, and this electronic shield that protects the Earth, life on Earth would end. Life on Earth would end. Thank God.

References

- University of Colorado at Boulder
-- Invisible shield found thousands of miles above Earth blocks 'killer electrons
- D. N. Baker, A. N. Jaynes, V. C. Hoxie, R. M. Thorne, J. C. Foster, X. Li, J. F. Fennell, J. R. Wygant, S. G. Kanekal, P. J. Erickson, W. Kurth, W. Li, Q. Ma, Q. Schiller, L. Blum, D. M. Malaspina, A. Gerrard, L. J. Lanzerotti. An impenetrable barrier to ultrarelativistic electrons in the Van Allen radiation belts. Nature, 27-11-2014.

Seas and waters

Earthquakes, storms waves and tsunamis in the deep ocean

It's a cold, dark, and terrifying environment filled with hurricanes, huge waves, turbulence, and storms
In new research published on 8-13-2014 in the journal Geophysical Research Letters, scientists were able to observe and record powerful storms and hurricanes occurring in the depths of the oceans and seas! The discovery came as a surprise to researchers at the University of Tasmania because they did not expect to find such tornadoes in the deep ocean.

Scientific ships monitor these waves

Days before this discovery, scientists from the University of Washington studied the depths of the ocean and discovered powerful waves, and analyzed these waves for the first time. The research was published in the journal Geophysical Research Letters and the researchers emphasized that it is the first time humans have seen such waves forming in the dark depths of the ocean.

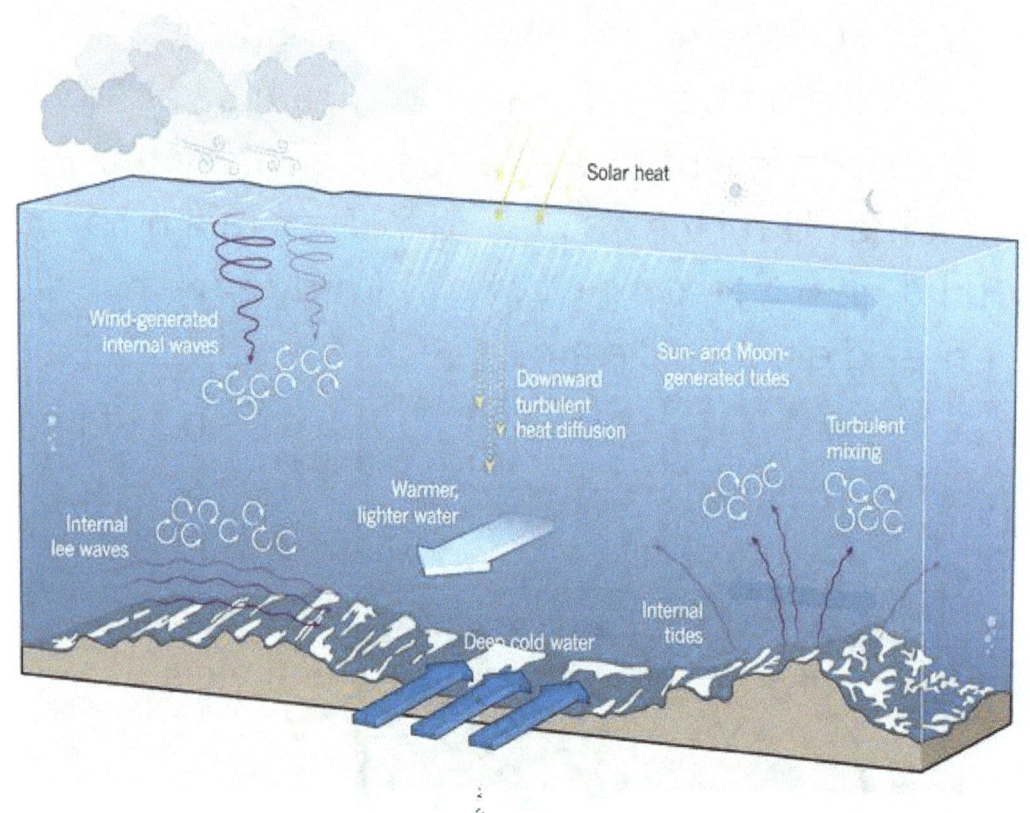

A blueprint for the importance of these deep waves and their role in mixing and sustaining ocean waters over thousands of years

A few years ago, scientists observed tsunamis generated by earthquakes in the deep ocean. Scientists emphasize that life under the deep waves is very difficult, darkness, storms and enormous water pressure ... Even modern submarines fail to stay long in such conditions and how many nuclear submarines have disappeared with their crews without anyone knowing anything about them until now.

This dark and difficult environment, which was only discovered in the 21st century, was mentioned in the Quran fourteen centuries ago.

References

- Seismic data tracks deep ocean storms
- abc.net.au/science/articles/2014/08/13
- oceannews.com/breaking-deep-sea-waves-reveal-mechanism-for-global-ocean-mixing
Deep Ocean Tsunami Detection Buoys

Freshwater reservoirs discovered on the ocean floor

Researchers from Flinders University estimate the amount of fresh water to be more than 500,000 cubic kilometers, enough for humans for hundreds of years Scientists have recently discovered huge amounts of fresh water stored beneath the ocean floor and flowing continuously. Scientists estimate the volume of these reservoirs to be half a million cubic kilometers of water, a hundred times the amount of water that humans have extracted from the water-bearing layers over the past century.

Fresh water has seeped beneath the ocean floors from several ice ages over the past 20,000 years. Huge amounts of water have formed reservoirs

---

References
-sci-news.com/geology/science-fresh-groundwater-reserves-ocean-
-Vast freshwater reserves discovered under the ocean floor which could supply future generations

Discovery of a sea beneath Earth's third layer

The Daily Mail published a strange discovery: An enormous sea 400-600 kilometers below the Earth's surface, under the third layer of the Earth's strata. This sea contains three times as much water as the Earth's surface... The discovery was made by scientists in March 2014.

And so we have a sea on Earth that we know, and right underneath that sea is a thin crust, and underneath that is the second layer of the Earth's layers, and then comes the third layer, which is a fiery layer from which volcanoes and molten lava flow. Underneath this layer is the newly discovered sea.

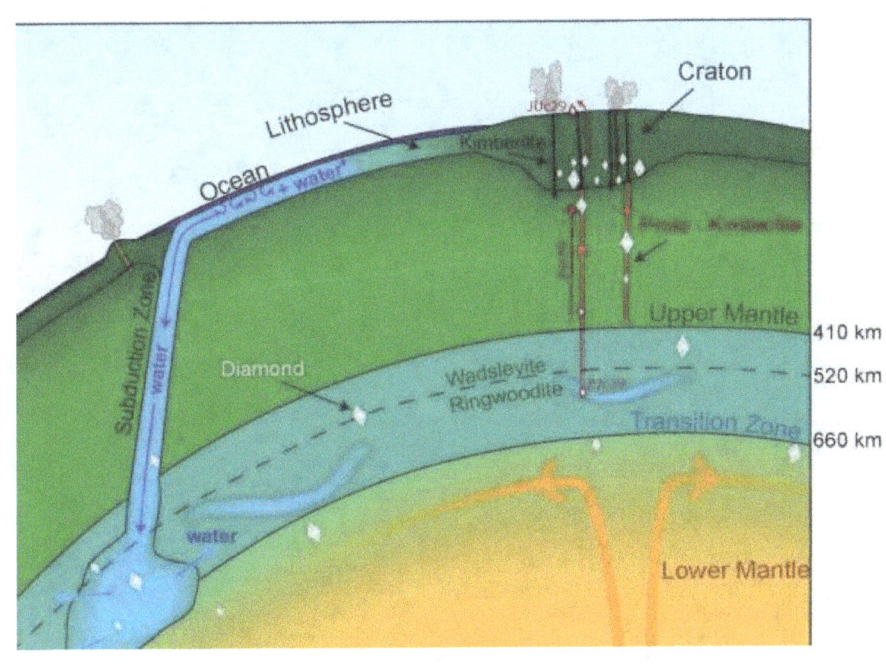

Photo source
DailyMail
2014/Mar

Thus, the number of Earth's layers becomes seven: 1- Earth's crust 2- lithosphere 3- upper mantle 4- middle mantle, which contains the newly discovered vast sea and is a transitional layer 5- lower mantle 6- outer core 7- inner core. So today the picture is clear and the number of layers of the earth has become seven layers. This fact is indicated by the Qur'an

So the scientific fact of the matter is: Beneath the sea is fire and molten lava flowing from the Earth's second layer, and beneath this fiery layer there is an enormous sea ... and we have the following order: Sea - Fire - Sea.

References

dailymail.co.uk/sciencetech/article-2579584//The-vast-reservoir-hidden-Earths-crust-holds

Sea foam

It's a beautiful phenomenon that can be strange to see for the first time, as sea foam can turn into a huge amount of foam that can stretch for tens of kilometers

This is a picture of sea foam, a complex phenomenon that occurs in all seas as a result of the intense mixing of impurities, organic matter, salts, dead plants and rotting fish... This leads to the formation of a very light foam, but it sometimes extends up to fifty kilometers.

Scientists explain the phenomenon as similar to putting a quantity of milk in a blender and blending it quickly to form a foam, which then dissipates into the air. The more violent the wave action, the larger and lighter the foam will be.

Some of the scientific facts about this butter include the following:

1- Scum is only formed in the case of rapid movement that occurs as a result of a hurricane or as a result of violent torrents. It always forms on the surface of the water at the top.

2- The weight of this foam or froth is very light and flies in the air like a vapor.

3- A small amount of water is enough to form a large amount of foam, which means that foam has no value, weight, or usefulness!

Treat your ailments with fasting

Try this method of treatment, especially for chronic diseases... Fasting... New studies come from renowned universities
 The University of Illinois at Chicago has released a promising research paper on the secrets of fasting. What caught my attention in this research is the duration of the fast, which is 16 hours, and allowing food for 8 hours (24 hours in total. This is similar to Islamic fasting). The study says that this method is the easiest of the fasting methods used in the West, and is the most acceptable to patients because it is easy to maintain
 Going without food and drink for 16 hours a day is much easier than the normal dieting technique of cutting calories at every meal... because few people stick to it... The 8:16 fast is easier in practice... This is what scientists from the University of Illinois say in a study conducted in 6-2018

This method of fasting was applied to a group of people and scientists found that it contributes to weight loss and improved blood sugar levels. The research results showed that the 16:8 fasting method, which means abstaining from food for 16 hours, contributed to reducing calorie intake by 300 calories per day, and weight loss by 3%.

This is a study from the University of Colorado Colorado will be launched in March 2018 to study the benefits of fasting after many studies confirming the same fact that fasting is a safe, cheap and easy way to treat excess weight and has other health benefits.

HEALTH

National Institutes of Health fund Colorado study of intermittent fasting

It would provide more reliable evidence of whether fasting is a safe and effective alternative to more standard methods of weight control.

Author: Jeremy Moore
Published: 7:40 PM MDT March 22, 2018
Updated: 3:11 PM MDT March 23, 2018

AURORA – Researchers at the University of Colorado Anschutz Medical Campus are looking for healthy adults who need to lose some weight.

They are planning the largest study yet of intermittent fasting.

It would provide more reliable evidence of whether fasting is a safe and effective alternative to more

The University of Washington School of Medicine has published studies on the importance of fasting and found that fasting has weight loss benefits as well. Researchers say: Animal studies have shown that reducing calories or reducing food consumption contributes to delaying aging, preventing heart disease, diabetes and cancer, and prolonging life.

However, fasting may have negative effects on some patients, so it is necessary to consult a specialist doctor, so God Almighty has given a license for this patient to break his fast during Ramadan.

One of the famous universities in the world, Harvard University, confirms that fasting has a good effect on the networks related to the generation of energy in cells, which leads to longevity.

Harvard study shows how intermittent fasting and manipulating mitochondrial networks may increase lifespan

Manipulating mitochondrial networks inside cells — either by dietary restriction or by genetic manipulation that mimics it — may increase lifespan and promote health, according to new research from Harvard T.H. Chan School of Public Health.

Researcher Heather Weir of Harvard University and head of the study says:

"Low-energy conditions such as dietary restriction and intermittent fasting have previously been shown to promote healthy aging."

References

- newatlas.com/intermittent-fasting-16-8-diet-science/55105/ 20-6-2018
- sciencedirect.com/science/article

Baby recognizes good and bad

A new study confirms that almost from birth, children are programmed to accept good and reject evil

A new study from the University of Chicago confirms that infants at a very early age can distinguish between good and bad behavior. The researchers of the study say they were surprised by the ability of young children to distinguish between good and bad behavior.

Professor Jean Decety, who led the study, says that according to the laws of evolution, children should be like each other without any "moral" differences. But the new study showed that children reacted differently to the perception of good or bad behavior... But what is truly amazing is the ability of these children, at an early age, to recognize, empathize with, focus on others and defer good behavior.

This purely completely contradicts the theory of evolution, which denies the moral aspect of the fetus and considers any moral habits to be library habits from the environment and society

This study completely contradicts the theory of evolution, which denies the moral aspect of the fetus and considers any moral habits to be library habits from the environment and society

---

References

How Parents Influence Early Moral Development, 29-9-2015

Fascinating Medical Facts

We offer our dear readers a collection of medical information and tips that everyone should know.

- In order to reduce food consumption, we advise you to chew your food well, try to talk to others between each bite and eat slowly and chew your food well, this technique will make you feel full and reduce the amount of food consumed.

- Before a meal, we recommend eating some salads and raw vegetables, as they contain a healthy amount of fiber that makes you feel full for longer and protects you from gaining weight.

- Be sure to eat two apples every day after washing well without peeling, as well as carrots, one of the best anti-LDL cholesterol substances, garlic, onions, soybeans and oat bran, and these foods have a great effect on the treatment of LDL cholesterol in the blood.

We advise diabetics to eat okra because the fiber in okra works to keep the blood sugar level within normal limits by controlling the amount of sugar absorbed from the intestines, and it is preferable notto fry okra but to eat it boiled.

Drugs make the brain eat itself

A new study warns against alcohol abuse, especially cocaine, finding that brain cells eat themselves while using the drug

A new study from Hopkins University conducted on the brains of rats shows that taking a large amount of drugs causes brain cells to kill themselves. Even when a pregnant mother takes drugs, the fetal brain slowly destroys itself.

Scientists noticed during their experiments that high doses of drugs make brain cells commit suicide, and the process of suicide takes place according to a process called autophagy This process occurs in healthy cells as a natural cleaning process to get rid of toxins.

This process does not lead to cell death in the normal state, but during heavy drug use, this process accelerates and breaks down the contents of the cell, which leads to its death... It's like the brain is killing itself.

Researchers say that the process of autophagy is necessary to clean the cell of accumulations and debris resulting from biological processes within the cell and harmful residues... a process necessary to keep the cell alive. But anesthetics make this process out of control.

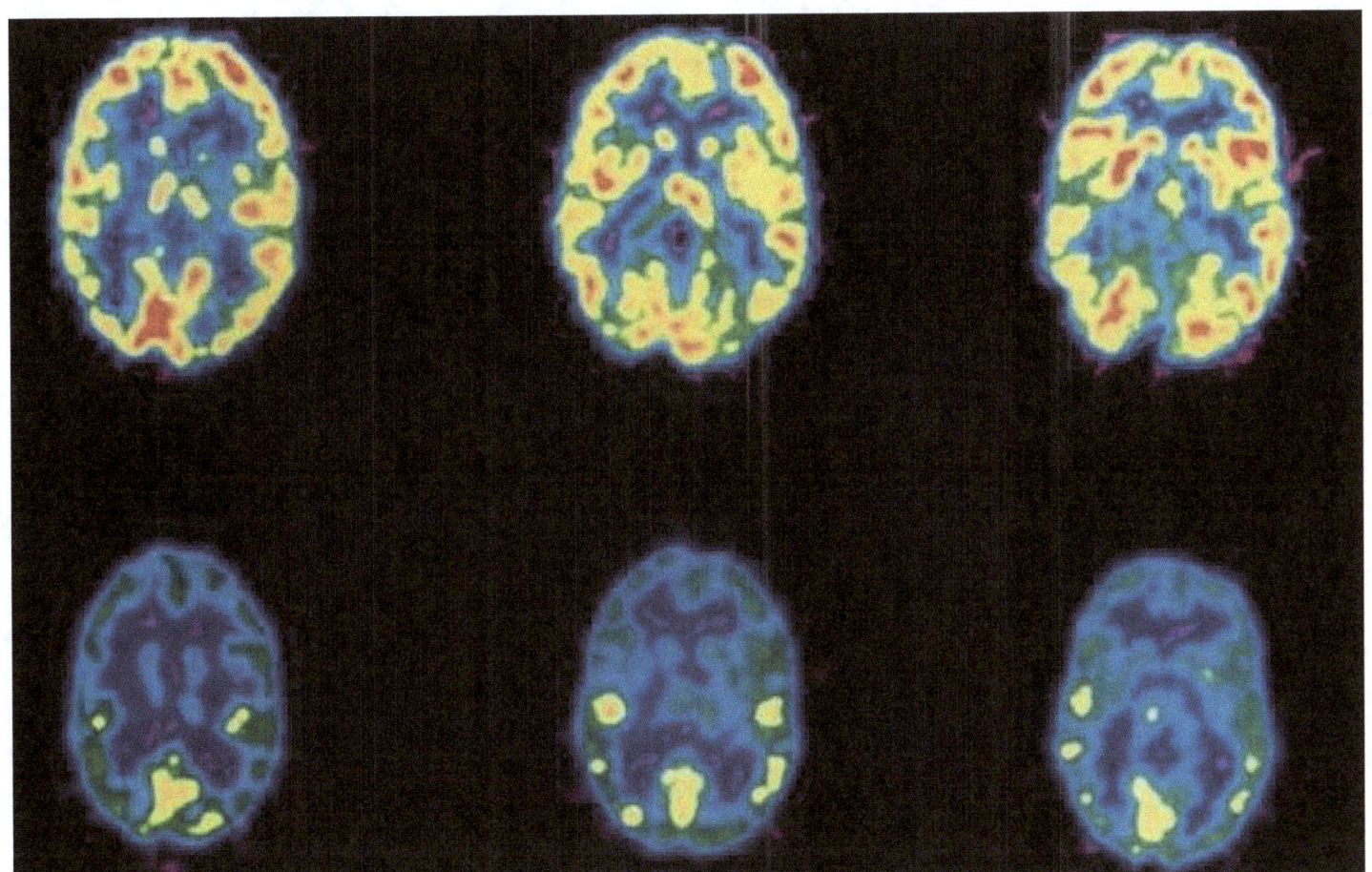

Magnetic Resonance Imaging (MRI) images show that the brain loses part of its cells during drug use, which is why scientists are strongly warning about the dangers of drugs and their devastating effects even on fetuses in their mothers' wombs.

References

hopkinsmedicine.org/news/media/releases/new_evidence_in_mice_that_cocaine_makes_brain_cells_cannibalize_themselves

Your brain makes a decision 7 seconds earlier

There are a lot of processes going on in our brain that we don't know about. Our decisions are not made in the moment, they are made some time in advance

In a study published in the journal Nature Neuroscience after observing the frontal region of the brain, or what is called the frontal cortex, which is the area behind the human forehead or what we call the nucleus, scientists found that this region is highly activated while thinking about making a decision.

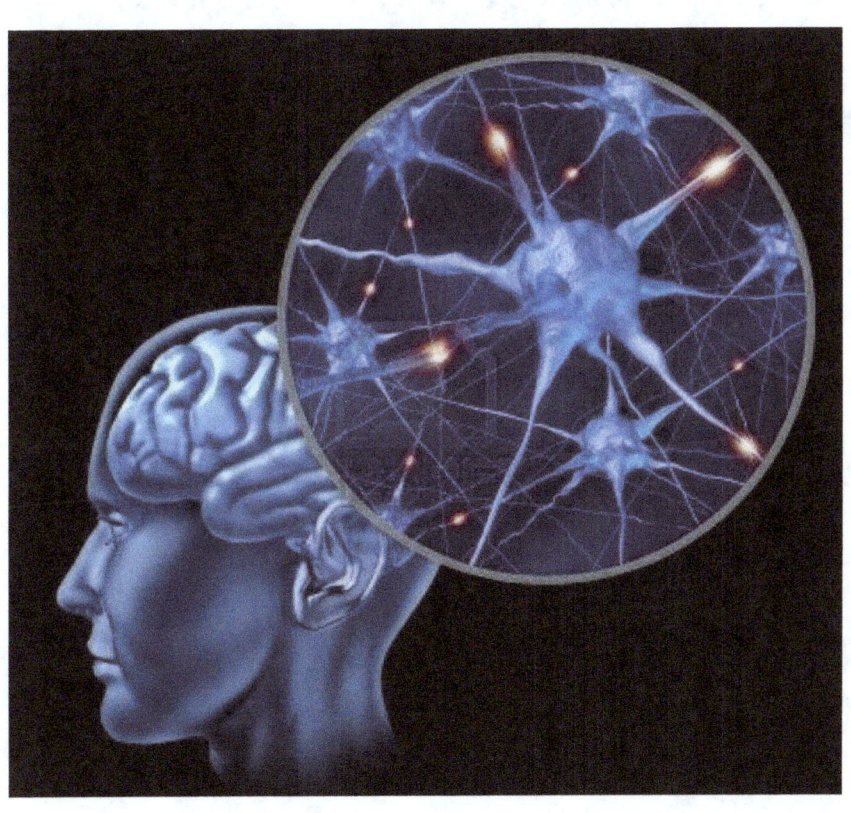

But surprisingly, the decision is made in the brain seven seconds before it is made by the human being. In an experiment using a magnetic resonance scanner, scientists were able to observe the nucleus region of the brain and found that when a person is faced with a set of options and must decide which one to choose, processes take place in the brain and a decision is made and after seven seconds this decision is determined by the person.

Referencess

Chun Siong Soon, Marcel Brass, Hans-Jochen Heinze & John-Dylan Haynes, "Unconscious Determinants of Free Decisions in the Human Brain." Nature Neuroscience, April 13th, 2008.

Sex scenes damage the brain

More than 20 scientific studies confirm that the damage caused by watching erotic sex scenes is similar to alcohol and drug addiction, and even more dangerous as it damages important parts of the brain

A new study conducted by researchers at Cambridge University found that the brain of a person who looks at taboos, especially pornography, behaves similarly to that of a person who is addicted to drugs and alcohol.

This study is the first of its kind (2013), where scientists used magnetic resonance imaging (MRI) of the brains of a group of young people addicted to watching sex movies, and the results of the study surprised scientists about the danger of sexual scenes and the need to limit their viewing!

Dr. Valerie Voon says, "We found tremendous activity in an area of the brain called the ventral striatum, which is the area responsible for reward, motivation and happiness.

Pornography addiction leads to same brain activity as alcoholism or drug abuse, study shows

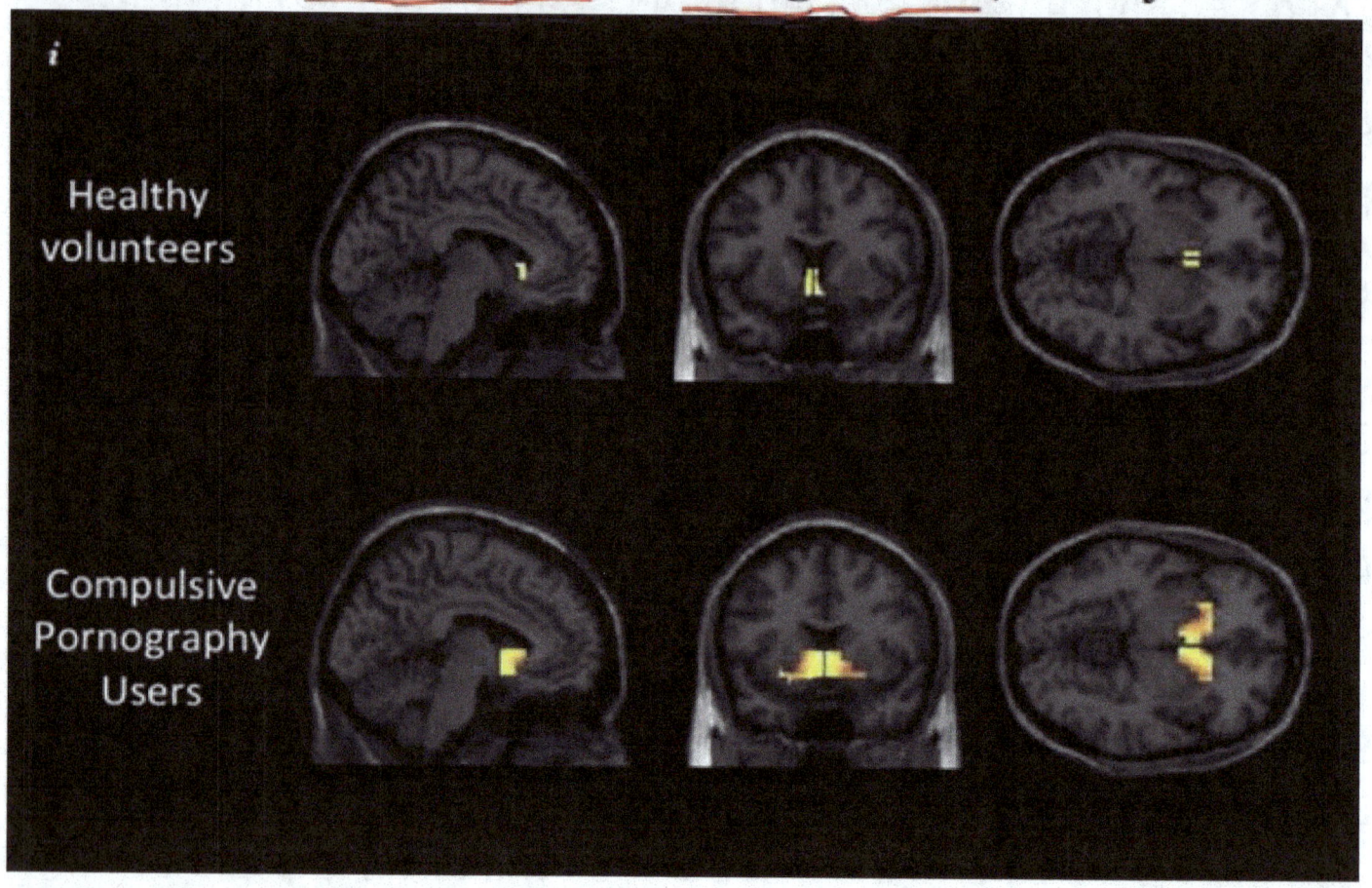

Cambridge University scientists reveal changes in brain for compulsive porn users which don't occur in those with no such habit

Magnetic scan images show that the area of the brain responsible for reward is unusually active during the viewing of pornographic scenes, and that the repetition of such scenes affects the brain in the same way that drugs and alcohol do

Marriage protects against cancer

New research conducted by scientists from Harvard University (September 2013) involving more than 750,000 cancer patients shows conclusive evidence that married people have a lower chance of developing various types of cancer.

The studyfound that marriage was more importantthan chemotherapy, curing 20 percent of cancer patients.

The Daily Mail confirms that since the 1990s, the number of unmarried people in Britain has increased, tripling in the last 17 years. Official records confirm that marriage rates continue to decline even today.

Researchers say that having a wife next to a man gives him a sense of confidence and provides him with adequate care and stability in life, so experts today recommend marriage as an appropriate way to live a disease-free and psychologically stable life.

The study showed that men were more likely to need marriage for cancer treatment, as more than 20 percent of men were completely cured, while 16 percent of women were cured of cancer due to marriage.

───────────────────

References

dailymail.co.uk/health/article-2430365/Marriage-improves-cancer-survival-rate

Honey: Endless benefits

- Treating cuts and scratches

Honey can be used to treat small cuts and scrapes. Honey is antibacterial and is one of the best natural remedies for wounds.

- Treatment of burns

Research has shown that honey has amazing benefits in treating burns, as it soothes the skin and promotes healing. Method of treatment: After cooling the burn with cold water or ice cubes, it is recommended to apply honey to the burn and bandage it with a bandage or plastic wrap, and change the bandage daily. The use of honey to cool burns is possible only when they are minor, and in cases of severe burns, it is necessary to go to a doctor.

- For stomach disorders

Stomach upsets can be treated with honey. Dissolve a spoonful of honey and a pinch of apple cider vinegar in a glass of water

This drink will rid your stomach of harmful gastric bacteria

- **To relieve throat pain**

Honey can help relieve a sore throat and soothe a cough. Method: Dissolve two tablespoons of honey in a cup of boiling water and add a little ginger, cinnamon and a spoonful of lemon juice.

- **To moisturize lips**

For those who suffer from dry lips, you can get rid of this issue in an easy and simple way, just brush your lips with a mixture of honey and a little coconut oil.

- **To treat fungal infections**

Some studies have shown that applying honey to the vaginal area can help kill vaginal yeast. It is recommended to apply honey, leave it on for 30 minutes, then wash it off with water and repeat once every 24 hours.

- **Honey is a natural skin cleanser**

Honey can be used as a way to cleanse the skin without expensive cleansing products. Honey is antibacterial, anti-inflammatory, and gentle on the skin, and is very suitable for those with dry skin and those with acne.

- **For silky hair**

Honey can be used as a natural hair rinse. It leaves hair feeling silky and soft. Method: After applying honey to the hair, leave it on for 10 minutes and then rinse thoroughly.

_____

References
dw.com

Dates and pomegranates are a powerful heart remedy

Fruit, especially the combination of pomegranates and dates, is considered a powerful medicine for many diseases, especially atherosclerosis, cholesterol, and high blood pressure.

A new study says that drinking pomegranate juice with a few dates is a very powerful treatment for heart disease. Pomegranates contain antioxidants that are essential for the body to minimize the oxidative processes that destroy cells. Dates also contain antioxidants that are necessary to eliminate bad cholesterol in the blood.

Dr. Fuhrman says on his website: **All fruits contain antioxidants, but the most powerful antioxidants are found in pomegranates.**

References
drfuhrman.com/library/article19.

Olive oil destroys cancer cells

In new research from New York City's Hunter College, scientists were stunned to find that olive oil kills cancer cells as quickly as 30 minutes!

The strange thing about this research is that the substance in olive oil, Oleocanthal, destroys cancer cells and preserves healthy cells.

In another study it was found that olive oil, especially Oleocanthal, helps prevent Alzheimer's disease, or the brain cell damage disease that causes memory loss, according to the University of Louisiana. Oleocanthal activates brain cells and may help them regulate their functioning, prolong their lifespan, and mitigate cell destruction.

References
news.rutgers.edu/research-news/ingredient-olive-oil-looks-promising-fight-against-cancer/2015

Treat yourself with green tea

A Japanese study has found that drinking green tea significantly reduces the risk of dying from various deadly diseases. Japanese researchers who conducted the study on more than 40,000 people found that drinking green tea reduced the risk of fatal cardiovascular disease by a quarter. However, British cardiologists said the benefits may be related to the quality of Japanese meals as a whole, which are considered healthier than those eaten by people in the West. The findings were published in the Journal of the Association of American Physicians.

References
news.bbc.co.uk/hi/arabic/sci_tech/
newsid_5340000/5340900

Walking nourishes the brain

One of the new discoveries is that walking every day for half an hour stimulates blood circulation and thus the brain gets enough blood to perform vital processes and daily activities to the fullest. This helps the brain maintain itself against dementia or premature aging.

Previous studies have found that walking is very beneficial for diabetics as well as helping to regulate blood pressure and lower triglycerides in the blood. Walking has benefits for treating back pain, nerve inflammation, vertebrae and preventing atherosclerosis...

Scientists say that the best type of walking is not too fastand not too slow, and this is called frugal walking

Learning in children

In a recent scientific research conducted by scientists that showed impressive results of the learning process in children, it was found that children are born with little information that helps them acquire additional information, unlike other creatures such as insects and animals.

A grasshopper, for example, is created with all the information needed to perform its tasks immediately, it doesn't need to learn to fly, it doesn't need someone to teach it to hunt, and it doesn't need to know enemy from friend. This knowledge is stored in its brain cells before it is even born!

But a child who is deprived of learning for some reason will have developmental, mental and cognitive issues. There are children who are born with an issue in some parts of the brain, and this issue leads to an inability to acquire new information, and thus the child develops autism or something else... and as a result, he is not a normal human being.t

Rather, human creativity, intelligence, and mental abilities depend on the amount of information acquired during childhood, so we can summarize the scientists' discovery with two points, and this is a new discovery that no one knew about before the advent of the 21st century:

1- The child is born with no knowledge except his ability to suckle from his mother's breast, and this is a very necessary piece of information that he acquired by sucking his fingers while in the womb

2- The sources of human learning are hearing and sight, while the brain processes and stores information, and when there is any issue with hearing or sight, the child develops abnormally and has less knowledge. For example, a deaf child cannot learn as well as a child who hears well.

DID YOU KNOW?

the information provided in the "Did You Know?" facts is based on widely accepted scientific, historical, and technological knowledge.

Did you know there are more stars in the universe than grains of sand on all the beaches on Earth?

Did you know black holes are so dense that even light cannot escape their gravitational pull?

Did you know Saturn's rings are made up of billions of pieces of ice, rock, and dust, ranging from tiny grains to house-sized chunks?

Did you know one day on Jupiter lasts only about 10 hours, despite the planet being the largest in our solar system?

Did you know Venus is the hottest planet in the solar system, even though Mercury is closer to the Sun?

Did you know Mars has the largest volcano in the solar system, called Olympus Mons, which is about three times the height of Mount Everest?

Did you know a neutron star is so dense that a sugar-cube-sized amount of its material would weigh about a billion tons on Earth?

Did you know Pluto was reclassified as a dwarf planet in 2006, but it has five known moons, the largest of which is Charon?

Did you know your skin is the largest organ in your body and makes up about 16% of your total body weight?

Did you know the human nose can detect more than 1 trillion different scents?

Did you know the average human body contains about 37.2 trillion cells?

Did you know the average adult human body has around 60,000 miles of blood vessels?

Did you know your brain generates enough electricity to power a small light bulb?

Did you know your body replaces about 98% of its atoms every year?

Did you know babies are born with 300 bones, but by adulthood, they have only 206 because some of the bones fuse together?

Did you know the average person produces about 25,000 quarts of saliva in a lifetime, which is enough to fill two swimming pools?

Did you know the muscles that control your eyes are the fastest muscles in your body, making them capable of moving in less than 1/100th of a second?

Did you know the shortest war in history lasted just 38 to 45 minutes between Britain and Zanzibar in 1896?

Did you know the Great Fire of London in 1666 destroyed 80% of the city but resulted in only six recorded deaths?

Did you know the Eiffel Tower can be 15 centimeters taller during the summer due to the expansion of iron in heat?

Did you know that Christoph Columbus was not the first to discover America?

Did you know ancient Egyptians used moldy bread to treat infections, using the natural antibiotic properties of penicillin?

Did you know Leonardo da Vinci could write with one hand while drawing with the other at the same time?

Did you know in ancient Rome, it was considered a sign of wealth to have long sleeves, as fabric was expensive?

Did you know a snail can sleep for up to three years if the weather conditions are unfavorable?

Did you know the heart of a blue whale is so large that a human could swim through its arteries?

Did you know a flamingo's pink color comes from the carotenoid pigments in their diet of algae and crustaceans?

Did you know an ostrich's eye is bigger than its brain?

Did you know dolphins have names for each other and can call out to specific individuals in their pods?

Did you know polar bears have black skin under their white fur, which helps them absorb and retain heat?

Did you know the first computer mouse was made of wood in 1964 by Doug Engelbart?

Did you know the world's first website, created by Tim Berners-Lee in 1991, is still online?

Did you know over 6,000 new computer viruses are created and released every month?

Did you know the word "robot" comes from a Czech word that means "forced labor" or "drudgery"?

Did you know the first 1GB hard drive, introduced in 1980, weighed over 500 pounds and cost $40,000?

Did you know more than 3.5 billion Google searches are made every day?

Did you know Earth's deepest point, the Mariana Trench, is deeper than Mount Everest is tall?

Did you know the Amazon Rainforest produces about 20% of the world's oxygen?

Did you know Antarctica is the largest desert in the world, even larger than the Sahara?

Did you know Russia spans 11 time zones, covering one-eighth of Earth's inhabited land?

Did you know the Pacific Ocean is larger than all the landmasses on Earth combined?

Did you know the Dead Sea is so salty that people can float effortlessly on its surface?

Did you know Africa is the only continent located in all four hemispheres: Northern, Southern, Eastern, and Western?

www.ingramcontent.com/pod-product-compliance
Lightning Source LLC
Chambersburg PA
CBHW062110220526
45471CB00010B/3685